Health Zone

W9-CFC-259

Stay Safe!

How YOU Can KEEP OUT OF HARM'S WAY

Sara Nelson

illustrations by Jack Desrocher

Consultant Sonja Green, MD

Lerner Publications Company
Minneapolis

For Sensei Kore Grate and the women and girls of Feminist Eclectic Martial Arts

All characters in this book are fictional and are not based on actual persons. The characters' stories are not based on actual events. Any similarities thereof are purely coincidental.

Lerner Publications Company
A division of Lerner Publishing Group, Inc.
241 First Avenue North
Minneapolis, MN 55401 U.S.A.

Website address: www.lernerbooks.com

Library of Congress Cataloging-in-Publication Data

Nelson, Sara Kirsten.
   Stay safe!: How you can keep out of harm's way / by Sara Nelson ; illustrated by Jack Desrocher.
        p.   cm. — (Health zone)
   Includes bibliographical references and index.
   ISBN 978-0-8225-7551-1 (lib. bdg. : alk. paper)
   1. Safety education—Juvenile literature. 2. Children and strangers—Juvenile literature. 3. Children—Crimes against—Prevention—Juvenile literature. 4. Child sexual abuse—Prevention—Juvenile literature. I. Title.
HQ770.7.N45 2009
613.6–dc22

                                                          2007049658

Manufactured in the United States of America
1 2 3 4 5 6 — BP — 14 13 12 11 10 09

Table of Contents

Tasha *hated* her family get-togethers.

She liked to see her grandparents and her cousins. She enjoyed talking to her aunts and uncles—at least most of them. *But one person made every get-together awful: Uncle Max.*

Max gave Tasha the creeps. She wanted to tell her mother how she felt, but she found it hard to explain the way Uncle Max treated Tasha made her feel uncomfortable. He always tried to greet her with a kiss on the lips. He asked her to dance and held her too close. Sometimes he would stand behind her and give her a back rub. That was supposed to feel good. But it didn't. It just felt strange. So Tasha didn't want to be around Uncle Max at all.

Tasha looked around the room. Sure enough, Uncle Max was slowly making his way toward her. She slipped out the door and headed to the bathroom. Tasha ducked inside, locked the door, and sat on the edge of the toilet seat. She wouldn't have to deal with him in here.

None of this makes any sense, Tasha thought. Her mother had always warned her not to talk to strangers. But Uncle Max wasn't a stranger. He'd known her since she was a baby. *How can someone in my own family make me feel so uncomfortable?*

Tasha wasn't sure what to do. She'd managed tricky situations before. When the class bully had been picking on her, she'd asked him to stop. When that hadn't worked, she'd gone to a teacher for help. That had solved the problem. But this problem was harder to talk about.

Tasha didn't feel comfortable telling Uncle Max how he made her feel. She would be too embarrassed. She would have to explain the situation to her mother. Meanwhile, she felt safe in the bathroom. Tasha flipped open her cell phone to check the time. Her family would be heading home soon. *She wouldn't have to wait very long.*

Chapter 1

WHAT IS Personal Safety?

Lots of us have felt the way Tasha does.

We know someone who makes us uncomfortable.
When we picture a threatening person, we tend to think about some scary-looking stranger jumping out of the bushes. It's true that we need to be cautious around strangers. But most victims actually know their attacker. In the end, the situation is what we really need to understand.

You need a number of tools to avoid trouble. Different people in different situations will need to use different tools. Fighting is rarely the solution to a problem. It's usually better to avoid a bad situation or to use your voice to avoid violence.

It may seem scary to think about attackers—whether it's someone we know or a stranger. But the point of talking about such people isn't to make you feel afraid. **The more you know about dangers, the better you'll be able to protect yourself.** *And that should make you feel safer.*

Sticking Up for YOU!

Personal safety is about knowing how to defend yourself. It's about learning to avoid danger. Sometimes staying safe means speaking up when something doesn't feel right. It can be hard to say that you feel uneasy. But look at it this way. Would you let anyone hurt your pet, a younger brother or sister, or your best friend? Of course not! Well, you're worth protecting too.

Boundaries and friends

Boundaries are important in all relationships. You set up boundaries with everyone, from your siblings to teachers to complete strangers. But what should you do when a good friend doesn't respect your boundaries? Maybe a friend always wants to copy your homework. Or maybe a friend wants to borrow your things and never gives them back. It can be hard to tell friends that their behavior isn't okay. But it's important to stand up for yourself. Clearly tell your friend what you want. Get help from an adult if you need it.

Why Do People Hurt Others?

It can be hard to understand why one person would want to hurt another. Some people may want money. Others might want to feel powerful or in control. Still others want to impress their friends. Sometimes people hurt others without even knowing it. They may think they are just playfully teasing. But the other person may not see it that way.

People can hurt others in many ways. Being able to recognize different kinds of threats can help you make decisions.

When Not to Trust

Don't trust someone if:

They tell you not to tell anyone about what they're doing.

They tell you to do something you know your parents wouldn't like.

They tell you to disobey your parents.

They tell you to do something illegal.

They threaten to hurt you in any way.

Bullying, Assault, and Abuse

Bullying is all too common in school hallways and playgrounds. It's also common on the Internet. Bullying can take many forms. A big kid might shove a smaller kid. A popular kid might say something nasty about a less popular kid. A bully might spread lies about another person. A bully might post nasty things about someone on the Internet. Or a bully might threaten to hurt someone.

Often, bullies do not act on their threats. But sometimes an **assault** occurs. An assault is a physical attack on another person. A mugger might hurt someone while stealing valuables. Or an argument between two people might end in an assault. Anyone who physically harms another person on purpose is committing an assault.

Abuse occurs when an attack—physical or verbal—comes from a person who has power over another person. Many cases of abuse happen within a family. This is called domestic abuse. Domestic abuse can be physical or verbal. When it's physical, it's often called domestic violence. It may be a husband beating a wife or a mother harming a child. And domestic violence often goes in cycles. If it happens once, it's likely to happen again.

Family trouble sometimes leads to kidnapping, or abduction. This is when a child is taken from his or her home. Most abductions happen when one divorced parent takes a child from the other parent. But strangers may also abduct children.

Amber Alert

In 1996, television and radio stations started broadcasting special bulletins designed to find kidnapped or missing children. The bulletins are AMBER alerts. The name stands for America's Missing: Broadcast Emergency Response. But it also honors a nine-year-old girl, Amber Hagerman. In 1996, Amber was abducted and killed while riding her bike near her home.

AMBER alerts go out whenever a child is

reported missing. Police inform local TV and radio stations about the missing child. The stations then broadcast information about the case. Pictures of the missing children appear on billboards.

AMBER alerts allow many people to hear about missing children. This way, people can be on the lookout for children who've been taken from their homes.

Pushing the Limits

Some threats aren't about shoving someone or calling a person names. Instead, they're about pushing limits. Uncle Max pushed the limits with Tasha. He touched her in ways that made her feel uneasy.

It's never okay for someone to touch you in a way that you don't like. If anyone does this to you, don't be ashamed to tell someone about it. You did nothing wrong. It's the person pushing the limits who should feel embarrassed.

Inappropriate touching can happen at home or outside the home. It can come from a stranger or from someone the victim knows and trusts. Religious leaders, teachers, coaches, family members, and babysitters are all people you trust. And usually they deserve your trust. But not always. **Remember— no one has the right to do something to you that you couldn't tell your parents about.**

Other kinds of inappropriate behavior don't involve touching. **Exhibitionism** is one such behavior. Exhibitionists expose their private parts to others. They enjoy the shock their victims show. **Peeping Toms, or voyeurs,** engage in yet another kind of behavior. These people peek through other people's windows. They look at people who don't know they're being watched.

More than half of the victims of these incidents never tell anyone. Sometimes they are afraid that nobody will believe them. Or they may worry that the person will hurt them again. A victim may even want to protect an attacker that is a friend or a relative. Some people don't trust the police. They may feel they can't prove they were attacked. Sometimes they're too ashamed. They think that maybe they caused the attack. *No victim ever asks for it.* **It can happen to anyone. If you report an attack, you might prevent the attacker from hurting someone else.**

Chapter 2

The Five Steps

Sometimes the world can seem like a scary place.

But you can learn how to protect yourself.

Many people, young and old, big and small, have used the tools described in this book to defend themselves. You, too, can use your mind and your body to stay safe. Professor Coleen Gragen, a martial arts expert from Oakland, California, helped to develop a five-step process to teach self-defense. (In this case, self-defense doesn't just mean fighting. It includes anything that helps people keep themselves safe while promoting peace.) The process is simple and easy to remember. You can use the steps in any order. You can even skip a few if you have to. *Different situations will require different strategies. Only you can decide which steps to use and in which order.*

Step One: Think Safety

The first step may be the most important step. It's also the most involved. (We'll explain more about it in chapter 3.) You need to know that safety is about awareness. Awareness includes setting boundaries, listening to your intuition (or gut feeling), and preventing problems.

Step Two: Voice

You can use your voice to tell others what you would like, to give commands, and to control a situation. Telling a bully that her comments make you feel bad may make her stop. Yelling can attract attention or make an attacker run away. You can even use your voice to set a boundary with a friend who isn't listening to what you need. **Don't be afraid to be heard. Your voice is a powerful tool to help you stay safe.**

Kindness as a Tool

You may not think that kindness could help you get out of a scary situation. But sometimes, kindness can be the best tool of all. Martial artist Terry Dobson illustrates this in a story. Dobson was riding the subway home from work one night when a man began shouting at people in the subway car. Dobson was ready to use his fighting skill to protect the other passengers. But then, an older man called the angry man over and began talking to him kindly. He told the angry man about the things he liked to do. He asked him questions. Soon, the angry man was crying in the older man's lap. He was no longer a threat. Dobson felt foolish. He had been ready to fight the angry man. The older man had made everyone safe, and he had used kindness to do it.

Step Three: Escape

The third step can help you get away from a situation quickly. There are lots of ways to escape. You could run away from a mugger. You could duck into a school or other building if you're being followed. Or you could go to a private place to avoid a situation. Tasha did this when she hid in the bathroom. Any of these steps can help you to avoid danger.

Hotlines

Check out these hotline numbers. They're great resources for reporting attacks or for just talking to someone about a situation.

National Center for Missing and Exploited Children: 1-800-843-5678

National Center for Victims of Crime: 1-800-FYI-CALL

National Child Abuse Hotline: 1-800-422-4453

National Domestic Violence Hotline: 1-800-799-7233

National Youth Crisis Hotline (Hope): 1-800-442-4673

Step four: Defend

What if using your voice is not enough and your attacker is going to hurt you? What if you can't escape? This step is for those situations. When all else fails, you might have to fight. Your body has five weapons to use against an attacker—head, hands, elbows, knees, and feet. Use them only when you have to. And remember to use only as much force as you need to escape or to stop the attack. As soon as you can get away, do it!

We'll talk more about defending yourself later in this book.

Step five: Tell

Getting help is just as important as any of the other four steps. Getting help may be as easy as shouting for an adult. Or it can be harder, like telling a parent about someone who makes you feel uneasy. Sometimes, getting help is just talking about an attack after it's over. *Talking about what you went through will help you heal.* And maybe, it will help police catch your attacker.

Chapter 3

THINK
Safety

Remember how we said in chapter 2 that the first step, Think Safety, is the most important one?

It's the most commonly used step and the one most people use to keep themselves safe.

Like Tasha, you have the ability to be aware of your surroundings. Have you ever felt someone come into a room, even though the door is behind you? We tend to think we're imagining things when we feel this. But we all have good intuition. You use your intuition when you know something isn't quite right. When Tasha realized that Uncle Max made her feel uneasy, she was listening to her intuition. It's easy to think it's just a funny feeling. But you shouldn't ignore it. **If you feel uncomfortable, it's time to take action.**

One way to take action is to set boundaries. How close do you let someone get to you? Do you let him or her touch you? Boundaries are going to be different for different people. You might hug your mom but not your soccer coach. Or maybe you hug your coach but don't want your neighbor to touch you at all. Boundaries can include personal questions as well. It might feel fine to give your friend your phone number. But you don't want to give it to a stranger. Only you can decide your boundaries with each person.

Have Information

Knowledge is power. Having as much information as possible is a big part of staying safe. Always know how to get in touch with a parent or guardian. And it's a good idea to have the phone number of another trusted adult to call if you can't get a hold of a parent or guardian.

Many neighborhoods have safe houses or shops set up for people who need help. Find out if your neighborhood has safe houses and know where they are. You can look for the McGruff sign too. The McGruff program puts pictures of McGruff, the crime-fighting dog, on windows of safe houses. The people at safe houses have been carefully screened. If you're in danger, they'll know where to get help.

Having information also includes knowing what to do in situations. Check out the "What If?" game included at the back of this book. Think about threats to your safety and how you would respond to them. What steps could you use? Which one would you use first? Being aware is all about having some ideas ready if you need to use them.

Think Safety with Strangers

Most kids know not to talk to strangers. If a stranger asks you for help, tell him you'll find another adult to help. You may think you're doing a simple favor by giving a stranger directions or helping him find a lost pet. But responsible adults don't ask children for help.

An adult who's trying to get you to do something dangerous will try to gain your trust. She might offer you money or a job. Sometimes she'll pretend to be a family friend or someone in charge. Check with an adult you trust before accepting a present from anyone—or letting anyone take your picture. Never get into a strange car or go to a stranger's home.

Be aware of what you tell a stranger.
It's easy to give too much information because you're trying to be friendly. It's not impolite to tell someone you don't give out certain kinds of information. You don't need to tell anyone your name, where you live, or who you live with. Strangers don't need to know which school you go to, which sports team you play for, or the names of your friends or family members. Keep this kind of information to yourself.

Think Safety on the Street

When you leave your house, you can do a few simple things to help you stay safe. First, make sure a trusted adult knows where you'll be, who you'll be with, and when you'll be back. Also be sure to introduce your parent or guardian to your friends and their families. It may be a little embarrassing. But that kind of information can be a big help if something bad happens.

Remember, there's safety in numbers. Groups of two are okay. Groups of three or four are even better. Stay in brightly lit and public places. Avoid bushes, dark corners, or isolated streets. Hang out with friends you trust. Sometimes you can find safety in a crowded area. An attacker might not come after you in a crowd.

Don't ...

walk down the street wearing headphones and blasting your favorite music. CD and MP3 players may make you a target for muggers. The music also makes it harder for you to be aware of your surroundings.

If you feel threatened, take action! You may need help but don't know anyone around. Look for a mother with children. She will usually be someone you can trust. Know how to use a cell phone or a pay phone. And don't forget that a cell phone with a dead battery is worthless.

When someone drops you off at home or at a friend's, ask the driver to wait until you get in the door before driving away. You can even arrange a signal. Turn and wave or flip the porch light on or off once you've decided that everything is all right.

When you're going someplace new, know where you're going. If you're using a bus, have your money ready. Choose an aisle seat so you can move easily. If someone bothers you, loudly tell the person to stop and then change seats. Get the other passengers' attention. The driver can help. On a train or subway, choose a car with a lot of people in it. Not many people are willing to attack you in a crowd.

The mall may be a public place, but it isn't always safe. People get distracted when they shop or hang out with friends. **Criminals know that.** If you need to use the bathroom, have your pals wait outside in a group. Girls who carry purses should hold onto them. That way, a criminal can't reach over the stall door and grab them. If you're getting picked up at the mall, be on time and wait in a brightly lit, public place. If you are ever uncomfortable or want someone to walk with you to your car or the bus stop, ask a security guard.

Gun Safety

Is there a gun in your home? If so, remind your parent or guardian to keep it unloaded and safely locked away. If you see a gun lying around—or if you ever see anyone pick up a gun—get out of the room and contact a trusted adult right away. And never, ever point a gun at another person. It doesn't matter whether you think the gun is loaded or empty. It's not worth the risk.

Think Safety at Home

Your home is one of the safest places you can be. But that doesn't mean that you don't have to worry about safety there. If you're going home and your mom, dad, or guardians aren't there, you can carry your key in a pocket that zips or buttons. Carry the key out of sight. That way, nobody will know that you're the only one home. If you get home and the door is open or a window has been broken, don't go inside. **Go to a safe house in your neighborhood and call 911.**

If you know that you're going to be home alone, it's okay to arrange for a trusted adult or older friend to stay with you. Have a phone number where your mom, dad, or guardians can be reached. And, for extra safety, you can get the number of a friend or neighbor. When you're home alone, you should lock your doors and windows. You can also turn on an outside light. Don't blast the stereo or TV, or you won't be able to hear the phone or doorbell.

And if the phone rings when you're home alone, *don't* tell the caller that you're alone. If the caller wants to speak to someone else, offer to take a message. Say that the other person can't come to the phone at this time. Never give out personal information on the phone. If a caller asks you a lot of questions, you can say that you'll have an adult call back to answer the questions later.

Have your mom, dad, or guardian tell you if they're expecting anyone to come by while they are gone. If the doorbell rings and you're not expecting anyone, you can yell, **"I got it!"** even if there's nobody else home. That way, the person on the other side of the door won't know you're alone. Look out a peephole or window before you open the door, and don't open the door to anyone you don't trust. **You can always tell the person (through the door) that your parents are not available and to come back later.** Just because someone says he or she is a plumber, postal worker, or salesperson, that doesn't mean that you should open the door. A delivery person can leave a package outside the door or a note with pickup instructions.

If people come to your door and want to use the phone, have them wait outside and offer to make the call for them. You can also direct them to the nearest shop or public phone. If they want to use the bathroom, send them to a public restroom nearby. You can also direct them to a neighbor in any of these situations. If someone is in trouble, you can call **911**, but don't open the door.

Think Safety at School

You can take steps to stay safe at school too. You may have to deal with a bully or even a pushy friend. Verbal or even physical abuse may be a problem. An even bigger danger is when a student brings a gun or another weapon to school. If you find out that someone has a gun, tell an adult right away. You may feel like you're being a tattletale. But this is a really dangerous situation. You need to take it seriously.

Most problems at school happen between classes, in hallways, and near lockers. If you don't dawdle, you should be able to avoid many issues. If you do have to face a bully, try using words to stop the bullying. You can also try to avoid the bully. Remember: treat everyone with kindness and respect. But you shouldn't be afraid to stand up for yourself. Talk to a trusted adult about any problems you have at school.

Projecting an Image

Think about how your body language affects how others see you. Imagine these two scenes.

①　You are walking down the street, head down, shoulders hunched, and arms crossed. You take small steps and look at your feet. If you meet someone coming the other way, you avoid making eye contact and you make sure you're not in the way.

②　You are walking down the street, head up, shoulders back, and arms swinging freely at your sides. You take long steps and are alert and aware of your surroundings. If you meet someone, you make eye contact and you may politely change your course. But you don't make yourself look small, and you take up the space you need.

If you were an attacker, which kid would you go after? The first one, right? The second kid seems like someone who can handle the situation. That's the image you want to project.

Think Safety on the Internet

The Internet is another place where you have to guard your personal information. Social websites like Facebook and MySpace are a lot of fun. But you also need to be careful. Anyone with Internet access can look at your profile unless you protect it with a password. If you give too much information, people can find out who you are and where you live. Don't post your full name, address, phone numbers, school name, or the town you live in. That information makes it easy for a stranger to find you. **And any stranger trying to find you probably isn't up to any good.**

Keep a parent or guardian in the loop on what you're doing online. Let them see your website and your blog. They can help you build a site that is safe but still shows off your personality.

If anyone you "meet" online wants to meet in person, you need to be very careful. This is risky. Some people use fake identities. They may not be who they say they are. Someone who says he's a boy may really be a man. Or a woman! Criminals often use the Internet to find victims. You can't be too cautious. If someone wants to meet you, tell a parent or guardian right away. If you're determined to meet an online contact, at least get a parent's or guardian's permission and bring an adult along. Meet the person in a public place. Don't say where you live. And don't ever leave the public place together.

Think Safety with Body Language

The way you carry yourself can have a big impact on your safety. Few attackers want to mess with people who can take care of themselves. Walk confidently and be alert. Even if you are small or scared, you can give the impression that it's a bad idea to attack you!

Eye contact is a big part of body language. **Making eye contact signals confidence.** If looking someone in the eye is difficult for you, look at the throat. That can make a person think twice about attacking.

The way you stand is another important part of body language. **Stand with your feet shoulder-width apart, bend your knees, and put one foot slightly forward.** You should be able to turn, run, and kick from this position. Keep your hands open and available (not in your pockets). But don't put up your fists. That looks as though you want to fight. Keep someone you don't trust at least two big steps away—more if the person is much bigger than you are or seems threatening. That way, you have the time and space to run or fight.

Chapter 4

When Thinking Safety Isn't Enough

You can avoid many scary situations just by being aware.

That's what makes that first step so important. But sometimes, awareness isn't enough. That's when the other steps come in.

Let's take a closer look at those steps.

Making Yourself Heard

Maybe, despite all your awareness, you have to deal with someone who is mean or angry or wants to hurt you. Move on to step 2—using your voice. You can use your voice to startle an attacker. You can use it to get help from bystanders. You can also use it to calm someone down and avoid a fight. (This strategy is called de-escalation.) Sometimes you can't calm an angry person. But if you're in danger, it's worth a try.

You can also use your voice to set a boundary. Saying "no" lets other people know what kind of behavior is and is not okay. It tells them when they have gone beyond our limits.

Saying "no" can be a tough thing to do—especially when we know and like a person. But it's important to let others know when they've made us uncomfortable. If another person is doing something that you don't like, don't pretend nothing is happening. Don't laugh it off. You want the other person to know that you're serious. Clearly explain how you feel. Make the person aware that the behavior is not acceptable to you.

"Back off!"

"Leave me alone!"

"Go Away!"

If you're in physical danger, be direct when telling someone what you need. Use statements that are clear and simple. Don't apologize for standing up for yourself. And don't be afraid to repeat yourself. You want to get your point across. **"Go away!"** or **"Back off!"** or **"Leave me alone!"** are all clear and simple statements. You can use any of these if you feel threatened. Depending on the situation, you might also try, **"This person is not my mother (or father)!"** or **"Call 911! This person is hurting me!"** When you know the abuser, tell the person clearly what you want. **"Please stop"** is simple and effective. Use a loud voice if you need to get people's attention. Even if no one is around to hear you, a deep, loud voice can stop an attack. Shouting can scare an attacker off. But if you think that your life would be in danger if you yelled, don't do it.

"No!"

Calling 911

If you're in danger or have been the victim of attack, call 911 for help. Be ready to tell the dispatcher what is happening and where you are. The dispatcher will stay with you on the line until help arrives. Remember, 911 is just for emergencies. Never call it just for fun. Someone else may really need help while you're wasting a dispatcher's time.

Getting Away

Escape is another way to get out of a bad situation. If you feel threatened, you can run. You can hide. You can cross the street or change your path to avoid someone. You can also get away when someone grabs you.

Escape can also mean changing your habits. If the attacker is someone you know and the abuse happens repeatedly, see if there's a pattern. Then change your routine to see if that helps the situation. For example, if the person confronts you whenever you're home alone, make sure you always have a friend with you. If it happens every Saturday night, spend Saturday night at a friend's house.

Escape strategies may be different when you don't know the attacker. If someone is following you, try switching directions. If someone is following you in a car, try turning around and going back the way you came. It takes time to turn a car around. So you may be able to escape. **Look for a store, a safe house, or a public place with lots of people.** Find a phone to call for help.

fight or flight

When humans and other animals
are faced with danger, they
have a reaction called fight or
flight. They either fight against
a danger or run away from it.
When the fight-or-flight response
kicks in, the body releases two
major chemicals. These chemicals
are adrenaline and cortisol.
Together they move blood to the
muscles, speed the heart rate,
and slow digestion. This gives
the body extra energy. When
the threat seems to be gone, the
body returns to its normal state.
The adrenaline and cortisol levels
decrease, the heart rate slows,
and digestion speeds back up.

If someone grabs your wrist, a simple way of escaping is to tuck your elbow into your body and whirl your hand and forearm in circles. If your attacker is really strong, you may need to hold one hand with the other and use it to help pull away. Don't interlock your fingers. You can break them that way. Sometimes a good kick to the knee or shins can distract the attacker long enough for you to make your circles and escape.

If someone grabs you around the neck, grab the attacker's pinkie fingers and bend them back. If you can't do that, try kicking or jabbing with your elbows.

It's rare for a stranger to grab someone. But you should know what to do just in case. That way, you'll be prepared to act.

Defending Yourself

If escaping from a situation doesn't work, try step 4. Your response will depend on the attacker, the situation, and what you can do. **Fighting is a last resort.** Never hurt someone out of anger. Use this step only when the other three steps haven't stopped your attacker or if you don't have time to use those three steps.

In defending yourself against an attacker who may be older, bigger, and stronger than you are, there aren't any rules. If you have done all you can to avoid a fight, then you need to do whatever is necessary to keep yourself safe. The thought of poking someone in the eye may gross you out. But doing it might keep you from being harmed. If you are fighting and one tool doesn't work, try something else. If someone wants to take something from you—your wallet, for example—it's better to just give it up. You can always replace your wallet. But you can't be replaced.

Heroes in kung fu movies use exciting, head-high kicks to fight off bad guys. But it takes years of practice to pull those off. For most people, they're not a realistic option. Instead, keep it simple. Read on to learn some easy ways to keep safe.

Martial Arts and Self-Defense

Taking a martial arts class has many benefits. But if you're just interested in staying safe, a self-defense class is better. Depending on the style, martial arts classes teach  punches, kicks, throws, and more. These moves take a long time and a lot of practice to master. Many of these moves would be difficult to use during a real attack, even if you had practiced for years. A good self-defense class will teach ways to prevent an attack before it becomes physical. It will also give you a few simple self-defense moves that will work without a lot of practice. Martial arts classes are great. But they're more about exercising the mind, body, and spirit than they are about self-defense.

Your Weapons

Just as there are five steps to safety, you have five reliable weapons on your body. As you may recall from chapter 2, those weapons are your head, hands, elbows, knees, and feet.

Your **head** is hard. It makes a good weapon if you use it to hit an attacker's face. Just avoid using your own face because that would hurt you.

Your **hands** are very useful as weapons. You can use your fingers to pinch, poke, scratch, and pull. A poke in an attacker's eye is one way to make a fast getaway. To make a proper fist, curl your fingers into your palm and hook your thumb over the outside. Make sure your thumb's not sticking out somewhere. Otherwise, you may break it.

Your **elbow** is hard and bony. That makes it an ideal weapon. If you're tall enough, you can elbow an attacker in the head or neck. If you can't reach that far, an elbow to someone's ribs might make the person pause.

Your **knees** are also hard. They can be a good tool to use—especially if the attacker is holding you very closely. With your knee, you can strike the groin or head of the attacker. To use your knees, raise your leg up quickly. Try to use as much force as possible to escape.

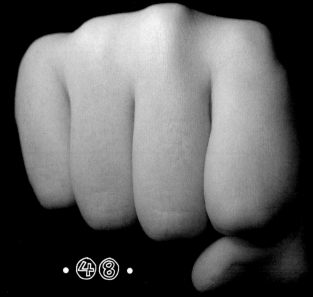

Kicking with your feet is another option. You should kick with the top (or instep), the outside edge (or blade), or heel of your foot. If you kick with your toes, you could break them. Kicking an attacker's shin or groin might allow you to escape and find help.

Remember that it's not hard to hurt someone.
You want to fight only when you feel you're in danger. Never fight because you're angry or you want revenge. You want to stay safe, not show off.

Talking about a Threat

It's not always easy to tell someone when you've been hurt or threatened. But speaking up is smart. A trusted adult can help you deal with a bad situation. He or she may have ideas to help keep you safe. This is the fifth step in staying safe.

If you've been attacked, tell an adult right away. The information you give can help stop your attacker from acting again. Try to remember all the details. Explain exactly what happened. It's also important to explain how you feel. If you don't have a parent or guardian to turn to, you can talk to a teacher or the school principal. Police officers, firefighters, counselors, and doctors are some other adults who can help you.

Sometimes it may not be enough to tell just one person. Adults aren't perfect, and they don't know everything. It's possible that the person you tell will not know how to help you. Caring adults want you to be safe. But sometimes people who've been abused themselves don't know how to help other victims. Or they may think that this is just how life is. The adult may also be afraid of your attacker. If this happens, **keep telling people until you find someone who can help.**

Conflict Resolution

Try using these steps to resolve conflicts.
They can work with friends, family,
classmates, or almost anyone.

1. Cool down.

2. Talk to the other person about the
   reason for the conflict.

3. Agree to resolve the conflict.

4. Have each person explain what
   happened, using "I" statements instead
   of "You" statements.

5. Brainstorm solutions.

6. Agree on a solution.

7. Act on the solution.

Describing Your Attacker

If you're the victim of a crime, try to pay attention to what your attacker looks like. The police will have a much better chance to catch your attacker if you can give them a good description. **Was the attacker tall or short?** What color hair? **Did the attacker have any scars, moles, tattoos, or other marks? What was the shape of the attacker's face?** Were the eyes close together or far apart? Any detail can help. If your attacker was driving a car or truck, try to remember the make, model, and color of the vehicle. Remembering the license plate number is even better.

Noticing these details can be hard if you're panicked. But you can practice paying attention to what people look like. If you're shopping, see if you can describe a cashier once you're out of the store. Can you picture the cashier's face? Could you give a description that would help someone else identify the cashier?

Saving Evidence

Sometimes victims want to get rid of all evidence of a crime. They want to forget the incident and put it behind them. They might throw away the clothes they were wearing when they were attacked. They might take a bath or a shower right away. Don't do this. It will destroy evidence the police can use to track down and put your attacker in jail. Instead, go to an emergency room and explain exactly what happened. Have the staff contact the police and be ready to describe your attacker. Ask hospital staff to take photographs of your injuries. They can help bring an attacker to justice.

Don't clean the scene of the attack either. **Leave everything the way it was and show a trusted adult.** If you noticed the attacker touching anything, remember this. The police may be able to get fingerprints.

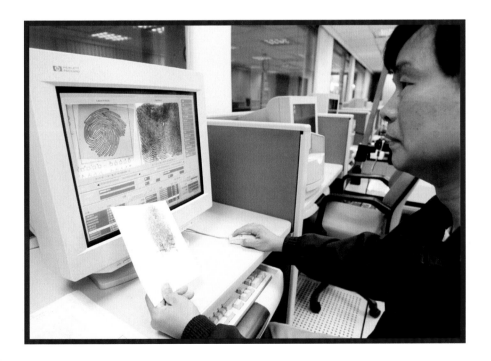

Healing

People who have been attacked or abused go through a lot of emotions. They may feel shock, denial, anger, shame, self-pity, guilt, sadness, fear, and anxiety. Sometimes victims have a hard time going to sleep, or they may wake up in the middle of the night. Some of them have nightmares. They can have a hard time concentrating or suffer from headaches or stomachaches. They may have a hard time eating. These are all normal reactions. Everyone deals with emotions differently. Someone may seem fine but still be affected by an incident. Sometimes people relive the incident or avoid anything that might make them remember it. Sometimes they don't want to talk to their friends anymore, or they may get startled easily. Talking to a counselor or therapist can be helpful in this case. Talking to a friend or trusted adult helps a lot too. *You're not alone.* **It's important to remember that.**

Be Aware—Not Afraid

It can be scary to think about attacks. Once you learn about all the bad things that happen, the world might seem like a pretty frightening place. But remember: attacks by strangers are rare. And there is help for people experiencing bullying or abuse. You should never have to go through a scary situation on your own.

Learn all you can about personal safety.

That way, you'll know how to protect yourself. Learning the five steps is a good start. They are clear, simple, and effective. Use them to be more aware of your surroundings and to

keep yourself safe.

Quiz

Now that you've read all about personal safety, try this quiz to see how much you know. Please record your answers on a separate sheet of paper. (Answers appear near the bottom of page 57.)

1. Most victims are attacked by:

a. a teacher

b. a stranger

c. someone they know

2. When using your voice to let someone know that their behavior makes you uncomfortable:

a. don't laugh it off, and clearly explain how you feel

b. pretend that nothing is happening. It's best to act as if the situation doesn't really bother you.

c. be really quiet

3. If you discover that you're being followed by a car, you should:

a. yell at the driver

b. turn around, go back the way you came, and look for a safe place to get help

c. freeze right where you are

4. A bully is picking on you at school. He is calling you names and threatening to beat you up. Which is the best way to deal with this situation?

a. punch him in the stomach and run away

b. ask him to stop, and tell a trusted adult if he doesn't

c. post nasty rumors about him on the Internet

5. **Many people who have been the victim of an attack don't report it because:**

  a. they feel there's no way they can prove they were attacked

  b. they're ashamed or embarrassed

  c. both a and b

6. **If you've been the victim of an attack, what should you do?**

  a. tell a trusted adult

  b. write about it on your blog

  c. find your attacker to get revenge

7. **Your mom is picking you up at the mall. Which is the best place to wait for her?**

  a. in the parking lot

  b. on the nearest street corner

  c. in a busy, brightly lit area

8. **If someone grabs your wrist:**

  a. beg the person to let you go

  b. tuck your elbows into your body and whirl your hand and forearm in circles

  c. spit in the person's face

9. **What should you do if someone in an Internet chat room wants to meet you?**

  a. tell a parent or guardian right away

  b. invite the person over to your house

  c. meet the person at a public place before going off alone together

10. **You're home alone and a stranger rings the doorbell. Which of the following should you not do?**

  a. yell, "I got it!" even though nobody else is home

  b. tell the stranger that you're home alone and are afraid to open the door

  c. tell the stranger through the door that your parent or guardian is unavailable and to come back later

1. c, 2. a, 3. b, 4. b, 5. c, 6. a, 7. c, 8. a, 9. b, 10. b

WHAT IF?

To stay safe, know how to react to different situations. Think about some possible situations and what you might do in them. You can play "What If?" with your parents, siblings, or friends. Or you can play it by yourself. Just imagine a threatening situation. Then decide how you'd act to keep safe. No right or wrong answers are given here—it's all about thinking of possibilities.

Check out the situations below. Go through the five steps. Which ones would you use? In what order would you use them? Think up your own situations as well.

- What if you were walking home from school one day when a van slowed down near you? Two men are inside, and they ask you to come over to their window to look at a map. They tell you that they're trying to find a street and they need your help.

- What if a friend was creating a website for you and wanted to include a picture of you doing a trick on your skateboard? The picture shows the front of your house with the house number clearly visible. Your friend also insists on including your cell phone number on the site.

- What if you were on a busy street, and a stranger grabbed you and tried to shove you into a car? Imagine the same situation on an empty street.

- What if the parent of a friend tried to touch you inappropriately?

- What if you got home from school before your parent or guardian and you noticed that the back window has been smashed open?

- What if a kid at school kept teasing you about the clothes you wear?

- What if three older kids came up to you and demanded money while you were riding your bike in the park?

Glossary

abuse: a physical or verbal attack, usually coming from a person who has power over another person

AMBER alert: a special bulletin designed to help find kidnapped or missing children

assault: a physical attack on another person

body language: the gestures and movements you use to communicate with others

de-escalation: a strategy in which people use their voices to calm others down

domestic abuse: abuse that happens within a family. Domestic abuse can be physical or verbal. When it's physical, it's domestic violence.

exhibitionism: a behavior in which people expose their private parts to others

fight or flight: a reaction to stress. The fight-or-flight reaction prepares people to either fight against a danger or to run away from it.

intuition: a "gut" feeling

kidnapping: taking a child from his or her home. Most kidnappings happen when a divorced parent takes a child from the other parent.

safe house: a home in which you can find help in an emergency. People who live in safe houses have been carefully screened. They are prepared to help those who need it.

voyeur: a person who peeks through other people's windows. Voyeurs are also called Peeping Toms.

Selected Bibliography

Department of Justice. *AMBER Alert: America's Missing: Broadcast Emergency Response.* 2007. http://www.amberalert.gov (November 26, 2007).

Dobson, Terry. "A Soft Answer." In *The Overlook Martial Arts Reader: Classic Writings on Philosophy and Technique.* Randy F. Nelson, ed. Woodstock, NY: Overlook Press, 1989.

Dyer, Gerri M., ed. *Protect Yourself and Your Family from Crime and Violence.* Rockville, MD: Safety Press, 1998.

Dyer, Gerri M, ed. *Safe, Smart and Self-Reliant: Personal Safety for Women and Children.* Rockville, MD: Safety Press, 1996.

Grate, Kore. Feminist Eclectic Martial Arts. Personal communication with author, 1993–2003.

Jacob Wetterling Foundation. *JWF.* N.d. http://www.jwf.org (November 26, 2007).

Lanoue, Nancy. Thousand Waves Martial Arts and Self-Defense Center. Personal communication with author, October 20, 2007.

McCallum, Paul. *The Parent's Guide to Teaching Self-Defense.* Cincinnati: Betterway Books, 1994.

National Center for Missing and Exploited Children. *Personal Safety for Children: A Guide for Parents.* N.d. http://www.ncjrs.gov/html/ojjdp/psc_english_02/intro.html (November 26, 2007).

Nemours Foundation. *KidsHealth.* 2007. http://www.kidshealth.org/parent/ (November 26, 2007).

Rape Abuse and Incest National Network. RAINN. 2006. http://www.rainn.org/statistics/index.html (November 26, 2007).

Chaiet, Donna. *Staying Safe at Home*. New York: Rosen Publishing Group, 1995. This title is written for girls. But it has plenty of useful information for boys as well. It focuses on safety in the home.

Chaiet, Donna. *Staying Safe on the Streets*. New York: Rosen Publishing Group, 1995. This title focuses on safety on the streets. It includes a discussion of stalking, being robbed, and physical self-defense. It's written for girls but has plenty of good information for boys as well.

Goedecke, Christopher J., and Rosmarie Hausherr. *Smart Moves: A Kid's Guide to Self-Defense*. New York: Simon and Schuster Books for Young Readers, 1995. The authors describe self-defense techniques, including punches, blocks, and kicks. This book includes sections on how to escape from holds and also emphasizes nonviolent skills.

Iedwab, Claudio, and Roxanne Standefer. *The Peaceful Way: A Children's Guide to the Traditions of the Martial Arts*. Rochester, VT: Destiny Books, 2001. This title discusses martial arts, their benefits, how they apply to self-defense, and how to use martial arts skills responsibly.

KidsHealth for Kids
http://kidshealth.org/kid
This health site features information on a variety of issues, including gun safety, dealing with bullies, staying home alone, and much more.

National Center for Missing and Exploited Children
http://www.missingkids.com
This site focuses on kidnapped children. It includes pictures and information about missing kids, safety tips for parents and kids, and a tip line.

NetSmartz
http://www.netsmartz.org
This website has good information about staying safe online. It has a section for both kids and teens, complete with interactive quizzes and quirky characters rapping about Internet safety.

Raatma, Lucia. *Safety around Strangers*. Chanhassen, MN: Child's World, 2005. *Safety around Strangers* gives readers tips about staying safe at home, on the street, and on the phone and Internet.

Photo/Illustration Acknowledgments

The images in this book are used with the permission of: © age fotostock/SuperStock, pp. 4, 13 (top), 56 (top); © BryceBridges.com/Alamy, p. 6; © Larry Williams/CORBIS, p. 9; © vario images GmbH & Co.KG/Alamy, pp. 10, 17; © Martin Ruetschi/Keystone/CORBIS, p. 11; © David McNew/Getty Images, p. 12; © Doug Plummer/Photographer's Choice/Getty Images, p. 13 (bottom); © Marina Jefferson/Taxi/Getty Images, p. 14; © Richard Lord/The Image Works, p. 15; © Steffen Schmidt/Keystone/CORBIS, p. 16; © John Eder/Stone/Getty Images, p. 18; © Mika/zefa/CORBIS, pp. 20, 24, 39; © Angela Hampton Picture Library/Alamy, pp. 21, 28, 58; © Jose Luis Pelaez/Iconica/Getty Images, p. 23; © Royalty-Free/CORBIS, p. 25; © Sean Cayton/The Image Works, p. 26; © Peter Alvey/Alamy, p. 33; © Jim West/Alamy, p. 34; © Erin Patrice O'Brien/Taxi/Getty Images, p. 37; © Chris Rout/Alamy, p. 38; © Barros & Barros/Photographer's Choice/Getty Images, p. 41; © Photodisc/Getty Images, p. 42; © altrendo images/Getty Images, p. 43; © Julie Caruso, p. 45; © Mario Tama/Getty Images, p. 46; © Adrian Green/Stone/Getty Images, p. 47; © Sami Sarkis/Photographer's Choice/Getty Images, p. 48; © Marc Debnam/Stone/Getty Images, p. 50; © Dick Blume/Syracuse Newspspers/The Image Works, p. 52; © Richard Chung/Reuters/C⬚ ⬚⬚⬚⬚ 54; © Charlie ⬚⬚⬚⬚⬚⬚⬚⬚et/Stone/ Getty Im⬚⬚

Front cov⬚

Abou⬚

Sara Ne⬚ ⬚ghteen
in Mont⬚ ⬚ more
than fou⬚ ⬚called
Wu Chie⬚ ⬚l Arts and
taught a⬚ ⬚rls during
much of⬚ ⬚sistant for
the Sch⬚ ⬚ountry
Institute⬚